Tray-Sean Ben Salmi aka Child Genius Advisor

10 SECONDS TO CHILD GENIUS

-

From Eton Road to Eton College™ Part 2

TRAY-SÉAN BEN SALMI
AKA
CHILD GENIUS ADVISOR,
AMAZON #1 BEST-SELLER, AWARD-WINNING AUTHOR
AS SEEN ON RADIO & TELEVISION

Published by Influencer Publishing in 2019

Third Edition

ISBN: 978-1-913310-10-3

'YOU' called this book to you because you live in an ever-changing world and you want to be a part of that change. Everything starts with an idea or concept in your mind. The idea or concept is given form by taking action and manifesting your **dreams**. Then the magic begins to happen during the construction process. That process begins here and **now**!

This book is dedicated to children and youth everywhere...

Cover Design by Swiss Graphics

ACKNOWLEDGEMENTS

I would like to thank God for all of my wealth, health, happiness, opportunities, abundance and support that I have received and experienced! I would also like to take this opportunity to thank all the amazing people that have supported my family and I, guided my family and I and taught my family and I along our journey. I really and truly would not be where I am today if it was not for family and friends.

I would also like to take this opportunity to acknowledge those of you who have delivered adversity to my life because this has made me stronger and inspired me to convert my adversities into empowerment. After all I am who I am today as a direct result of everything I experienced to date.

I would like to acknowledge my older sister Lashai for coming up with the title of this book: FROM ETON ROAD TO ETON COLLEGE

CONTENTS

Imagine how you'll feel as you finish reading the last page of this book.

Foreword

Did you know that youth unemployment is one of today's biggest global challenges?

Did you know that youth entrepreneurship offers a solution to the above, because innovative solutions for economic growth among young people is on the increase?

First of all, I would like to identify Tray-Sean Ben Salmi as a 14-year-old teenager that I hold in high regard, not because he's my son but because I truly admire and respect him upon so many levels in the context of business and personal matters. Tray-Sean is a remarkable teenage genius, who boosts a host of wisdom, skills, experience, products, services, unique opportunities and accolades etc. Who I have found to be passionately supportive of the work of others, devoid of triviality, and nobody's fool.

Tray-Sean Ben Salmi, prior to this book, had already made a lasting impact in both academics and his personal development. Would you agree that few teens do in our current climate of media coverage of our teens? He participated in Child Genius and was recognised as 1 out of 20 of the smartest children in the UK for 2017, he then went on to co-author 10 Seconds to Child Genius™ with Philip Chan and became 1 of 34 selected to sit papers at the prestigious Eton college for Boys. For more information may I suggest that you purchase a copy of his books called I'M THAT KID – I'M

THAT KID - EMPOWERING YOU TO STEP INTO YOUR POWER FORMULA™ and 10 Seconds to Child Genius – The Road to Child Genius™. At the age of 6 years old Tray-Sean launched his first short story competition in school. Tray-Sean was 7 years old when he co-authored Kidz That Dream Big with his older sister Lashai Ben Salmi and the two of them would attend a host of prestigious investment, business development, branding and personal development events. At the time they were the only children in the room. At the age of 11 Tray-Sean was awarded an UnLtd award to run his first event called: I'm That KID – Bridging The Gap Between Fathers & Sons at Arsenal F.C. which provided a unique opportunity for fathers, their sons, Tray-Sean and his step-father and a fundamental resetting of the moral compass. Thanks to Tray-Sean's (and his four siblings) work, in large part, the direction of youth entrepreneurship now is beginning to point away from simply setting up business and towards helping children and young people to activate their sacred gifts, network building, have fun, dream big and make money doing what they love while gaining key life skills, experience and unique opportunities in abundance. As Tray-Sean's mantra says, "I'm That KID – As long as I believe in myself, I can be, do and have anything I desire". This, I completely agree to be true because whatever a man thinks of most, that is exactly what he shall create.

This book is exceptional and most clearly deserves to have a very wide readership, I think it should be carefully read by every child aged 5yrs plus both locally and globally. Quite frankly, nobody else could have written this book reason being that the author himself has consistently excelled in academics and having found meaning in his life with a host of mind-blowing achievements to date. What distinguishes this book is the fact that Tray-Sean is incredibly well connected, well informed, and a personal determination shines through every word; and yet Tray-Sean writes with the beauty of poetry. The book is so well written, it actually does deserve the worn cliché that 'I couldn't put the book down'.

Tray-Sean is an old soul in a young body, that is in affect way ahead of his time.

Tray-Sean experienced an extremely traumatic birth, which almost cost him his life. He has faced life threatening raspatory issues since birth which resulted in frequent visits to A&E and has been cared for by several consultants since birth. Tray-Sean was diagnosed with Asthma at the age of 4, the precise context I am referring to the above is to highlight what a 14 year old can do, not what he or she cannot do despite being seriously ill from birth and had to overcome anxiety and low self-esteem.

I appreciate the sincerity, consistency, commitment, service to others, creativity, courage, passion, compassion, empathy, determination and genuineness of Tray-Sean. This book is written in a 'friendly' style, I especially recommend this book to our friends, family, colleagues and all establishments who work with children and young people. Unfortunately, our teens face a host of challenges as they transition from childhood to adulthood this endeavour should never be at the expense of the current generation of young people. I believe that the content of this book with inspire children, youth and adults alike to achieve the same or better because Tray-Sean is 'keeping it real' when it comes to highlighting what could be possible when children and youth are allowed to Dream Big as young as possible. We should all be on guard when it comes to protecting our children's natural ability to Dream Big, self-belief and experiential learning.

This book will be invaluable to anyone who wishes to learn about aspects of fully embodying their dream. I know of no other source which matches the quality and breadth of writing as in the book and the other books that Tray-Sean has authored, in a brilliant and unique way; including young onset raw creativity, personal development, self-realisation and experiential learning. And yet the book does not paralyse any approach that you may currently be honouring. I recommend to you his unique chapter on Consequences of Delayed Creativity, for example. Tray's book, overall, takes the field of Youth Entrepreneurship much further forward.

In fact, the book brings up topics that I really wish I was exposed to during my own childhood, because I believe that I could have made much more

progress in my life by now. In the areas of developing my core life skills, financial literacy, personal development etc as early as possible. The book further broaches additional topics in a unique, incredible way so as not to make you feel uncomfortable.
One of my favourite parts of the book is About the Author, because this is where Tray-Sean shares what he has achieved to date. I believe that this will inspire others about what is possible when you choose to Dream Big, invest time in personal development and have fun learning.

I feel that through this book Tray-Sean will ultimately reach out to millions, whether they are academic, artistic and/or entrepreneurial. A word I have often heard used about Tray-Sean is 'Inspirational'. In today's society of modern-day materialism in the media, it is not uncommon for this word to be used, but in Tray's case it is richly deserved. Trays acts a focus for acknowledgement for values and attitudes which are right about this world, in generosity and warmth of spirit; of being educated, and great fun to be with. I know that family means a lot to Tray, words cannot possibly convey how proud we are of him for choosing to step into his power and embrace life despite being bullied, having low self-esteem, experiencing panic attacks due to anxiety brought on as a result of health challenges (life threatening asthma attacks).
This book is highly original. I think it makes the weather on so many key topics, such as the natural abilities of children being just as capable as adults

when it comes to entrepreneurship. Whilst it may make some people feel uncomfortable to embrace youth entrepreneurship, including 'experts' and people in the media. They need not just to hear but to listen carefully to the immense lifelong benefits of youth entrepreneurship and self-exploration. The book somehow combines being timeless, placeless, and yet firmly relevant to us all, in the here and now, and clearly ahead of its time. This inherent child/youth entrepreneurship culture does not particularly worry me, because of the sheer brilliance shared within this book.

I am honoured to be Tray-Sean's mother. He is the second eldest of five siblings and I must say that he inspires me and teaches me so much about life, myself and others upon a daily basis. Inside this book you'll be exposed to a host of inspiring content that could transform your life and that of children and young people around the world.

Therefore, I truly believe that this book is a step in the right direction.

Sabrina Ben Salmi BSc
Co-Founding Director of Harris Invictus, Multiple award-winning Mentor and Author, Speaker, Founder of Shift Happens, Dreaming Big Together - Mamas Secret Recipe™ and co-founder of The Conscious Entrepreneur Blueprint™

INTRODUCTION

I wrote this book to encourage you to go on and live your dreams; no matter where you came from: class, race, geographical location, heritage, gender etc.

This book can be used as a constant reminder that it's not where you start in life, however it's about where you choose to finish.

I desire to share my journey with you, with the aim to encourage you to fulfil your dreams and to just always go and get whatever you want in life. To always remind you that you can do, be and have anything that you desire in life as long as you believe in yourself.

I have chosen to be vulnerable with you by sharing my ups and downs to show you how you can achieve the same and even better in your own life.

ETON ROAD

I grew up on Eton Road, Ilford IG1, we moved there on 24/9/2007 and moved out 25/2/2017. I lived there with my mum, step-father and 5 siblings (my grandmother also lived with us for a couple of years until she had to become a carer for my grate grandmother). We had a spacious open plan 3-bedroom house with a massive garden. The interior was painted in magnolia with chocolate brown skirting boards and banister and we had chocolate brown carpet throughout our home. Once you entered our home you'd see our staircase that led upstairs to a huge bathroom which housed our toilet, shower and bath. Then my two sisters shared a room, my parents room was next and then the boys room which I shared with my younger brother. I come from a large family, therefore you can imagen that we experienced a host of ups and downs just like any other family. I really love being a part of a big family because there is always something to do and someone to do it with. We are all very different in our own unique way. Nan was the one who would always splash out on expensive designer items and would always get us out of trouble when we did something wrong. One of the things that my nan would often say is "Leave them, they didn't do nothing wrong – they're just kids. Are you okay my Tu-tu-luts/darling/push/du-duts?". If I am honest nan would say this all the time even when we did something wrong and I would just smile because mum wouldn't buy it, even if she would have to listen to nan out of respect. My step-dad was strict and enjoyed cooking. Mum spent a lot of time on personal development and raising us, she was easy going and it took her many years to speak her truth. Now that bring me on to Lashai, she's the one to watch because she was the one really in charge. We all listened to Lashai and she always knew how to get what she wanted no matter what. I'm the second sibling in line and I was just quiet and learned through observation. Yasmine is a loving and caring character and would always teach us to love more. Paolo is fearless and would always push the boundaries even when we told him not to and then there is little Amire. As the saying goes, size doesn't matter, I am beginning to believe that that saying was coined because of my little bro because he is totally unaware of limits, boundaries and fears. Therefore, I am sure that you could imagine what our day to day life was life ha, ha, ha, ha…

Downstairs was our open planned lounge and dining room that led to our galley kitchen. Our dining room backed onto our garden. In our garden we were able to fit so many things in it we had a trampoline, electric cars, football table, swing set, a see-saw and so much more. It was probably where we spent most of our time when it wasn't raining. We played there, we had parties there, we would have family dinners/BBQs during summer. In our house there was always that smell of either strawberries or the smell of cinnamon. Whilst growing up we didn't get to socialise much locally because the neighbours kept themselves to themselves. I can recall having a couple of sleepover at my best friend's (Alfie) home. Where we lived our neighbours were very unfriendly, and when we played football in our garden if we kicked the ball over (which happened a lot) they would either chuck our ball in their shed or they would pop our ball. Eventually we stopped playing football in our garden to prevent that from happening.

My Nan (Mary Paul) always wanted us to
Look our best and was always making sure that we had designer clothing and suits. I really appreciate all that my mum and grandma have done and continue to do for us.

I attended a local school in Redbridge. Unfortunately, I didn't have the best of experiences in primary school because I was often bullied. For example, in one of our lessons pencil lead was put in to my ear and it hurt a lot, things like this would happen upon a daily basis – if I tried to defend myself I would get into trouble. So, I decided to ignore them and just focus on completing my school work so I could go into and remain in top set, eventually I was put onto the gifted and talented list.

Can you remember a time when you had to pick your battles so you could move towards your desired outcomes?

I would often cry in school because I would frequently be targeted just because I was smart. I realised that I had to make friends with some of the popular kids in order to survive and I could help them with their maths. I also became good at football, which really helped to make me popular. Finally, I grew to know a lot of people who would always look out for me and who would protect both myself and my siblings if anything was to happen.

I was chosen to join our school football team and it was such a remarkable feeling to be on our school team. I have always loved to play football come wind, rain, snow or sleet – nothing could get in the way of me playing football. It was so much fun because we would often win. Outside of school I took football seriously by training daily to make sure that I stay fit and strong to ensure that I can always perform at my best.

I am a big believer of the saying that my eldest sister taught me 'practice makes improvement'

Would you agree that it is always a good idea to practice whatever you desire to master?

I had a traumatic birth and ever since then I have had respiratory problems. I was always in and out of the hospital so much so that I was known by the ambulance operators, ambulance paramedics and hospital staff. As a result, I would miss a lot of school. The good thing is I always stayed in top set because studying has always been something that I enjoy.

What do you enjoy doing?

What do you spend most of your time practicing?

Just know, that whatever you spend time on and focus on the most you will eventually master it and then go on to develop your own unique artistry.

The question is what will you become a master of as a result of what you have been doing daily?

Spending quality time together has always been one of our core family values. This meant a host of trips to Tunisia, Portugal and Chessington World of Adventures. LEGOLAND, Under 17 Car Driving, London Eye and London Dungeons. The River Themes Clipper boat, hosting workshops, speaking on stage, Greenwich Park, Science Museum, Barbican, Ice Skating, featuring in TV shows, newspapers, radio, writing books and Swimming. Unique opportunities and gifts, Horse Riding, Natural History Museum, British Museum and the British Library to name a few. I looked forward to every weekend because we always had so much fun spending time together as a family. One of my fondest memories is when my mum took us away to Portugal for my sisters birthday. We had so much fun in Lisbon with friends (Barbosa Family: Barbra, husband, Guilherme and Margarida), we stayed in their amazing four-story home. We had an amazing experience there as it was a private gated complex with a swimming pool, football court, basketball court, park, gym, football table etc. we would wake up at the crack of dawn and played until the last one was standing. We had so many friends and even if some of us couldn't speak the same language, that didn't stop us from having fun.

Can you remember a time when you had so much fun and didn't want the night to end?

My childhood has been and still is pretty amazing and at times I just have to express gratitude and pinch myself.

Whilst doing all of these amazing things both myself and my older sister (Lashai), were doing work experience with **Miles Fryer** and **Sammy Blindell** in their branding company called '**How to Build Your Brand**'. We worked with them for two years. We were also fortunate to be surrounded by a host of coaches, mentors and trainings. The interesting thing is, I selected Business Enterprise as a GCSE option and on the first day we had to do an assessment and my teacher was blown away by my extensive knowledge within branding, business, social media, media and marketing. I was allocated an expected grade of A* and he asked where I acquired my extensive knowledge and I proudly replied that I've been doing work experience with our family friends called **Miles Fryer** and **Sammy Blindell** in their branding company called '**How to Build Your Brand**'. The teacher contacted my mother to commend me and I am pleased to say that she was touched by my teachers excitement towards me. I also managed to join a football team call Redbridge FC thanks to Chris (Alfie's dad). Alfie was small and the other players would make fun of him pre-match – Alfie and I would just laugh because he was on fire once he touched the ball because his dad trained him as soon as he could walk. We were both stickers and we would enjoy being silent and allowing our actions to speak loudly because we had great chemistry within our team we were at the top of our league, and we lost very few matches. I can remember a time when I was the star of the match. One of my favourite memories was when my Redbridge FC coach came to our home to collect me for training, due to an Arsenal F.C. scout coming to watch me. There were a few times that he had attended, but I wasn't there because I was unwell and had difficulties breathing so couldn't attend training. My coached informed my mum and I that the scout liked me and the words he used was 'Tay-Sean is a born athlete; anyone can be taught technical skills - but athletes are born'. When I was well enough to attend training and match days the Arsenal F.C scout returned and was watching me play, unfortunately it wasn't one of my best games because I had lost my favourite football boots in transit to the match, I had to play in my spare boots that were meant for indoors. These boots didn't have much grip, so I kept slipping and sliding all over the place!

Can you remember a time when you wished that you had a better plan B?

How do you tend to react or respond when things don't go according to plan?

I must say that I felt devastated when I wasn't scouted, but I chose to focus on building my dreams no matter what.

BELIEVE IN YOUR DREAMS

I have always had many dreams and aspirations that I wanted to aspire to achieve when I grew up. However, when we spent time with our family friends (***the Unstoppable Family: Rhonda, Brian and Hanalei Swan***) they told me that it's not about what you want to be when your older, it's about what you choose to do now. It's about what you choose to do with your time now because tomorrow isn't guaranteed.

The question is what do you want to be now?

I love to play sports such as: football, basketball, handball, badminton and many more. I also enjoy mathematics, in my spare time I like to practice my mental maths and I like to go over GCSE test questions to make sure that I stay on top; so I can consistently push myself to do my best and always give it my all.

I also love to play football because I really love to use my mathematics to work out ways to improve my in game performance and different strategies to get pass a team's defence using my brain to analyse the way they play and find out a way to get past them. When I am on the field, I treat it like a maths problem.

Between the ages of around 9-11 I played football for REDBRIDGE FC. We were an amazing team and we had great chemistry; we were in division 1. I can remember our winning streak of around 9 games. I was man of the match for 4 of those matches and I scored at least 2 goals in each match.

At the time of writing my dreams and aspirations down I didn't know *'how'* I'd achieve them, but I had a huge *'why'* and then amazingly unexplainable things began to happen to help manifest my dreams.

The way I look at things in life is simple really...

1) Never give up
2) Come up with solutions
3) Always strive for the best
4) Focus
5) Relationships are key

1. When trying to achieve anything in life you might sometimes be knocked down and may feel like giving up, but you must always persevere no matter what. Most of the time you are so close to achieving your goal and then you give up, because it seems like it is no longer achievable. If you want new results you have to do something different. 'YOU' must want change, nothing will happen if your friends or family want you to change without you wanting to change.

2. In life when the Set targets and give each one a completion deadline to make sure that you stay on track.

3. You have to commit 100%: There are a lot of options and they can distract your focus, you must always strive to perform at your best no matter what and you should never be satisfied with your performance until you know that you have given it your all. 'YOU' must choose to commit to what's right for 'YOU'. What you are committed to and what you are interested in are two completely different things.

4. Stay Focused: Develop laser beam focus for your **dreams** and aspirations. There is no right or wrong way because there are pros and cons attached to each and every choice you make and be patient - little by little bit by bit. Only take part in activities and conversations that will bring you closer to having **FUN**, **DREAMING BIG** and making money doing what you **LOVE**.

5. Build strong relationships: Feed off positive input and spend quality time with people who encourage, inspire and motivate you. Stand on the shoulders of giants and 'YOU' will observe more, because you will become like the three people who you spend most of your time with. So, choose wisely.

I BELIEVE IN YOU AND I THINK THAT IT IS ABOUT TIME, THAT YOU START TO BELIEVE IN YOURSELF NOW

REVISION

When revising a subject, I usually enjoy watching videos, talking to mentors within that specific area of study and I usually test myself in a number of ways. When needed, I setup a study group of up to 4 people at the most. I prefer to study in small groups because it allows me to maximise how much I am able to learn as a pose to being in a large group where it would take longer for people to grasp all of the different concepts.

The key things I would advise that you do to help you focus when revising is either listening to music that calms you, I tend to listen to classical music when I am studying. You could create flash cards and/or write points onto post-it notes. I suggest that you try several techniques to enable you to choose one or a few that work best with your own personal learning style.

I find that when I revise for tests there are key features that I always use such as: Finding out as soon as possible when the test is. I usually find out at least 3-7 weeks in advance so that I have enough time to see what I need to go over . This allows me to allocate adequate time needed to revise for the test. Then I either get a mock test online or ask the teacher to provide me with one to see which areas will be covered in the test and this is also an opportunities for me to establish my strengths and weaknesses within the topic area. Once I have all of the above, I ensure that I gain adequate sleep, eat well and drink plenty of water. I would then get into a group of up 4 people to maximise how much we could learn together and then we could then teach one another new techniques of how to solve different questions. This is when I would create a timetable for both group and solo revision.

I always revise in the evening during the week and 12 noon during the weekend, because it is the time when my brain works at its best. I advise you to establish your golden time, it's different for each of us, you may be more focused during the morning, afternoon or evening – whatever works for you. I also make sure that I include frequent breaks to stop myself from overworking. Having frequent breaks will help you to stay on top of what you are doing; it helps you to analyse what you are doing and make the most out of you revision time. During your breaks you may desire to have a snack, meditate, relax, dance while listening to music, exercise, go for a walk or cycle do whatever feels right for you.

Another key tip that many underestimate is rewarding themselves for what they get right instead they spend too much time beating themselves up for what they get wrong. I used to beat myself up until I learned how to redirect my energy and thought processes. Be sure to celebrate all wins and simply learn from the feedback each time you get it wrong. Otherwise you will only get in your own way of progress. Be sure to make notes.

Here's some food for thought:

What is my best spot for revising?

Who do I tend to study best with?

What resources do I need to help me maximise my revision?

What do I need to stop doing?

What do I need to start doing?

What do I need to do more of?

What do I need to do less of?

What support mechanisms do I need to put in place?

What are the best ways to maximise my revision time?

When are the best times for me to revise and for how long?

What are my three core revision goals?

What is my big why in life and how is this a step in the right direction?

I am a true believer of the saying 'The bigger the WHY! The easier the HOW!'. Laser beam focus is absolute paramount when you have a burning desire that you wish to achieve, without focus and aligned action – how will you accomplish your desires?

ADAPTING & EXPANDING

When I was growing up I had to be able to socialise within professional, entrepreneurial, academic and investment networks. As a family we had to be open to new opportunities and we had to be fairly flexible and have faith that it would all work out in the end.

There were often times when we would have to journey back and forth from Heathrow at 6am to get to events, this was often the case for events across London. As a family we are so committed to our personal and collective development that we've also flown out to Bali, Indonesia for training with our family mentors Regan Hillyer and Juanpa Barahona. Since the age of 7, I've been taught to be fluid in life because in my capacity to be adaptable, then I will be able to expand. I have had to be willing to step outside of my comfort zone in order to learn new things. After all, every new experience resides outside our current understanding. Therefore, in life we have to be willing to become the version of us that we need to be in order for us to experience our desired outcomes.

To what extent are you willing to go to in order to commit to your personal development?

Are you open to being flexible and adaptable, in order to allow yourself to expand?

While at events my siblings and I would sell our products and services for pocket money. I must say that at first it felt uncomfortable as it was taking me outside of my comfort zone. However, it eventually became such a fulfilling feeling each time I sold a product or service – it made me feel significant and incredibly powerful.

I really appreciate how our mother brought us up, for example my mum used to give each of us pocket money up until a certain age. Then my mother decided that we should learn how to make money for ourselves. We had to learn how to find a problem, come up with a solution and then package that solution into a product and service spectrum. Hence why we created a system, workshops, wrote our books etc. We sell them to generate multiple streams of passive income. At first, this came as a huge shock to my system as I didn't understand at the time; but now I can see the benefits of doing so. As a direct result of the above, my siblings and I have gained a host of key life skills such as hot to pitch a product, how to entice to customer to invest, how to speak from stage, how to create a franchise, how to network with purity of intention, how to turn our passions into profits, creativity, how to focus on the benefits, kindness, paying it forward, respect for ourselves and others, how to take people on a customer journey etc to name a few.

Whilst going to all of these amazing events we got to meet some extraordinarily phenomenal people who have remained in our lives and continue to support us on our journey to this very day. We had to expand our network and awareness to allow more opportunities to come in. This allowed us to make space for better things that would not only have a positive impact on us but also it would go on to touch the lives of many others.

These events that we attend helped us to adapt, expand, gain confidence, activate our sacred gifts, inspired us to take aligned action and it created opportunities for us to socialise. Socialising was very important to us as a family. In our local area we were often very isolated because where we lived our neighbours often kept themselves to themselves therefore by attending these events helped us to gain key communication skills.

COACHES & MENTORS

As the saying goes

"When the student is ready the teachers shall appear" Chinese Proverb

This is one of my favourite proverbs, whenever I hear this proverbs it makes me think about what is my teach-ability index, and how open am I to change?

Here's the thing, until you are open to change no one is able to teach you anything, because you will not allow the information to enter. Therefore, the teacher will not show up until the person is ready to accept the guidance.

I have had many coaches and mentors for different areas of my life over the years:

Cheryl Chapman, Andy Harrington, Ewan Wong, Ash Domah, Juergen Pallien, Harun Rabbani, Tosin Ogunnusi, Sammy Blindell, Gary Hayes, Miles Fryer, Majid Khan, Malik Muhammed, Ivor Nickles, Trevor Carter, Harry Singha, Harry Price, Kris Kemery-Toone, Irene Soares, Mahmood Shaikh, Chris Cohen, Robert G Allen, Regan Anne Hillyer, Debbie Stoute, Chris Nash, and Juan Pablo Barahona to name a few.

During my early childhood my family and I realised the importance of mentorship and how having a coach/mentor would help us to put us on the fast track to achieve our goals and aspirations in life.

This is because success leaves clues. A coach/mentor gives steps to success, systems, blueprints, hidden secrets and that is why us as a family prefer to invest in coaches and mentors instead of just attending events.

Going directly to a coach/mentor/expert empowers you to achieve the same or even better. Throughout my childhood I had many amazing and profound mentors and coaches.

I have met wonderful and inspirational people at events, during the brakes we would do some networking which assist us to develop some amazing new contacts. Which would often go on to create new and unique opportunities.

During my early teenage years, I met a man called Phillip Chan and he told me to apply to participate in Child Genius, I managed to be recognised as 1 out of 20 of the smartest children in the UK. Following the show we decided to work together for a short period of time to co-author a book called 10 Seconds To Child Genius - we came up with our book idea in two hours, wrote the book in four days and became Amazon #1 Best Selling authors with in 2 hours and 30mins of our book going live. Philip arranged for us to host a workshop at a church of one of his friends. Philip helped us to edit all five of our first books. My mum arranged for us to be interviewed on London Live, I arranged for us to be featured in a couple newspapers and my mum arranged for us to work in partnership with Metro Bank. My mum helped to organise our workshop at a magnificent hotel in Canary Wharf. Philip invited me and my family to a networking vent to met George Ross (former adviser to Donald Trump) and we spoke on stage alongside one another a couple of times. While at this event, my mum connected with Juergen Pallien and his wife Linda.

There's one quote that my mum often shared with us and I will never forget it "You can lead a horse to water, but you cannot make it drink". My mum frequently used this quote to remind us that she can only do her best, but if we are not open to change then there is no one in the world who can make us experience anything else.

Where in your life are you stopping yourself from achieving better things?

Where in life are you your own barrier?

Where in your life are you limiting yourself to how much you can achieve in life?

What would need to happen in order for you to hire a coach/mentor?

ATTITUDE

I truly believe that your attitude will determine your altitude. What I mean by that is if you show a good attitude in life, then you will fly high and people will enjoy being around you and can go on to JV (Joint Venture) and create unique opportunities for each other.

Get into the habit of winning, find your talents and passion and have fun doing it, smile, be honest, be trustworthy. Expect unexpected pleasantries in abundance, ask for support when needed, be vulnerable, make others feel that there's something worthwhile in them. Set good intentions, do the right thing even when you think no one is watching, be teachable, value and appreciate yourself and others, set healthy boundaries, humble yourself so you can be of service to others and always convey gratitude for yourself, health, opportunities, other people and life.

For example, there was a boy at school who was always rude to the pupils and the teachers, he would always play tricks on them and do mean things and then one day he needed help, however no one believed him due to his poor attitude. There was another boy at school who made an effort to build positive relationships with teaching staff and his peers. One day he needed help, and both his peers and teaching staff were more than happy to offer their support. Therefore, always remember - how you conduct yourself in life matters and how you make others feel matters too. Hence why your attitude will determine altitude in life.

In my view gratitude is a key principle in life and learning the art of gratitude is absolute paramount. As, it will then help you to attract so many more things in life.

The clothes you wear carry a vibrational frequency. In life there are different dress codes for example there are those who wear suits, tracksuits and jeans etc. These determine the type of people that you are more likely to attract into your life. The clothes you wear have an impact on your attitude.

Never underestimate the power of your words. Your words have a substantial impact on yourself, people, and opportunities you are able to attract into your life. The way you speak allows people to make a rough interpretation of what type of person you are both physically and vibrationally. Never – ever underestimate your vibrational language, it is essential in attracting the right people and experiences to you.

In life embodying good manners and etiquette are very important as it will ultimately determine the way you conduct yourself around others.

Your outward bearing or your way of behaving towards others is also very important. Making a positive first impression is vital in life because a first impression can either make or break you. A first impression has an infringement on how the person feels towards you and if they feel like they should continue to build a relationship and/or help you out in any way. For instance, if you wanted to get a celebrity to endorse your brand and let's say someone was a close friend with that person, they'd give their opinion of you based on the impression you had previously given them.

Never give up and always remember that if you want something you should take aligned action to obtain it. Ask yourself these following questions:

What things should I keep to myself and which parts should I share?

Who am I sharing important information with?

Where and when should I share certain things?

Do the people who I am sharing these things with share the same values as me?

What type of people are you surrounding yourself with?

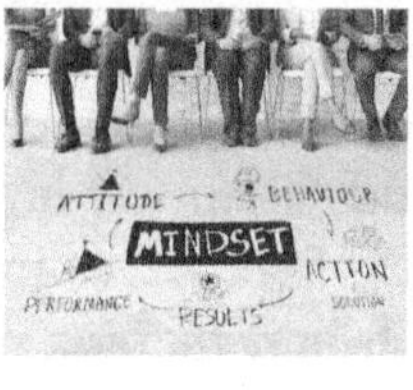

LIVING THE DREAM

I'm That KID and I AM living the DREAM!..

Whilst growing up my family always taught me that anything is possible as long as I my mind to it and that nothing in life was outside of my reach.

As a family we have achieved many extraordinary things such as traveling to Bali (on an all-expenses paid) when my sister won a $44k scholarship, being surrounded by world class coaches and mentors. Launching my very own short story competition at the age of 6. I co-authored my first book with my 11-year-old sister, when I was 7. I was honoured to receive a grant from UnLtd and was then offered the opportunity to speak on their stage after launching my I'm That KID brand at the age of 11. I've hosted my very first event at Arsenal F.C. (my favourite team). I am an award-winning author and an Amazon #1 bestselling author of 6 books (and many more to come). I am an award-winning public speaker and I have spoken on stage at The Best You Expo (the biggest personal development expo in the world). I have witnessed first-hand that thoughts become things, after all a man will achieve whatever his mind can conceive. I've participated in Channel 4 Child Genius. I have created a host multiple streams of passive income. I am an international coach and mentor, I have a business partner, we have worked in partnership with Mentro Bank etc. I have sat papers at the prestigious Eton

College. I've appeared in newspapers, radio, TV and so much more. I've also participated in brand campaigns for Legoland, Sony, Nicktoons, Primark and so much more as a former member of TEAM TROUBLE.

One of our family friends Gary Hayes (he's practically our adopted grandfather) plays a huge role in our lives and guides us to the right direction in so many ways. On a whole, I am really appreciative to have such an amazing role model to look up to.

I was honoured to have been given the chance by Sunil Chuni to speak at Virgin Money Lounge for the first ever Black History Month event. I was flabbergasted because I got the chance to speak on the same stage as my mother (Sabrina Ben Salmi) and my older sister (Lashai Ben Salmi). On the whole I was so amazed that I was selected to take part in such an important not to mention historical event. I delivered an emotional speech about my emotions pertaining to my father/step-father that brought me and the audience to tears.

Whilst growing up and to date my entire family often feature in newspapers, radio and magazines. We also enjoy doing as much as possible to support our community both locally and internationally. We are very passionate about helping each other, in as many ways as possible and feel overwhelmed with gratitude for the abundance of blessings that we are fortunate to experience and share with others.

My older sister Lashai Ben Salmi (aged 16 at the time) was gifted a scholarship from Reagan Anne Hillyer International. My mother, sister and I flew out to Bali, to take part in the Be Your Brand event. This was a truly transformational event; we learned a host of life changing teachings to take us and our brands to the next level and touch more lives on a larger scale. Upon our return myself, mum, Lashai, Yasmine, Paolo and Amire published books. We are forever thankful for being gifted such a life changing experience. It was truly an amazing experience to get a glimpse of what is available to us so long as I believe in myself I can be, do and have anything I desire in life. And the best thing is that Regan Hillyer and her partner Juanpa remain our family mentors to date. H.D.I.G.A.B.T.T?... (How Does It Get Any Better Than This?)

Believe

in yourself

&

you will be

Unstoppable

Around the age of 12 I attended Y.L.S (Youth Leadership Summit) for teens with my sister Lashai and Shane. Gordan Which is founded by a lovely man called Harry Singha (founder of Harry Singha Foundation). We were there for a duration of 4 days with experts from around the world teaching us life skills such as forgiveness, gratitude, paying it forward, financial education, entrepreneurship and so much more.

"I'M THAT KID"
I BELIEVE IN MYSELF
AND AS LONG AS I
BELIEVE IN MYSELF
I CAN DO, BE AND
HAVE
ANYTHING!

As a result of attending that event we maintained contact with Harry Singha. During a conversation Harry Singha asked me what I wanted to become when I grew up, I shared with him my dream of becoming a professional footballer. Harry, then introduced me to one of his friends Harry Price who is the co-founder (with his brother) of UK FOOTBALL TRIALS. I then attended the trials for FREE. Unfortunately, I didn't play my best on the day because I had a cold which prevented me from performing well. I am so grateful for the opportunity and I am looking forward to going back.

When I was around 8, I met a lovely lady called Cheryl Chapman (we call her auntie Cheryl). Cheryl encouraged me to do a board breaking session at one of Andy Harrington's events called Power to Achieve. Ever since then she has been a mentor in my life. She mentors and guides me onto the right path in life, to ensure that I stay on the right track. She is always there to support me to overcome obstacles in my life. Thanks to Cheryl, me, our mum, Lashai, Yasmine, Paolo and Amire are all public speakers.

One of our family dreams is to do a family world tour called Dreaming Big Together. We dream of sharing our vision with families around the world

What are your dreams and aspirations?

What support mechanisms do you need to help you to start living your dreams and aspirations?

THE POWER OF APPRECIATION

I am constantly thanking God for the fact that he is always looking over me and making sure that both myself, family and friends are safe and well.

I affirm that, I am the young man that I am today because of all of all the blessings in my life. I have been rushed to hospital in an ambulance more times than I can remember. Each time, I have been lucky enough to come out alive, that is all thanks to God for looking over me which has taught me to take my health more seriously. I appreciate the fact that I am still alive whole, healthy and complete.

My mum always taught us the importance of respect and appreciation. She would often say that it doesn't cost anything to be respectful and appreciative, but it could cost everything in life if you choose not to be respectful and appreciative. I have grown to realise that without respect and appreciation, you will be unable to achieve anything in life. It is important to build healthy relationships with people and respect and appreciation is an excellent starting point. For example, if you desire something and didn't have any relationships with people you would have to work hard on your own in the hope of attaining it. Whereas, if you have healthy relationships with people, one of your connections may have a close relationship with the person (or access to that which you desire to achieve) who could help you to achieve your goal quicker. Just like what happened for me with Harry Singha and Harry Price.

MUSIC

Did you know that ***MUSIC*** appeals to every emotion? It soothes and excites you, uplifts and inspires you. It moves you to happiness and reduces you to tears because music speaks directly to your hearts.

- Can you imagine life without music?

- Have you ever wondered how different genre of music influences your mood or actions?

- What is your favourite genre of music and how does it make you feel when you listen to it?

- Observe your friends, family and other people you know. Observe the music they listen to and how do you think that genre of music influences their mood or behaviour?

- Have you ever noticed how some music makes you feel playful, aggressive or even sad?

Music is a beautiful gift from God and it is one that people around the world treasure. Music is available to millions of people across the world at our finger tips. An abundance of music can be recorded, downloaded and played on a wide range of devices that can be slipped into your pocket. So much has changed since your parents were young. Please be very careful of the genre of music you listen to – listen to the words very, very, very carefully and become aware of the emotions that different genres of music elicit in you. Music, effects your emotions and unconscious mind in ways that you couldn't imagine. I am 14, so I know that this may sound much easier said than done. What I would say is that all new habits of behaviour take time, discipline and the desire to change. You are not alone; I am a mentor and offer a product and service spectrum to help you in your journey. So please do not hesitate to get in touch.

When revising You should choose your music carefully and wisely because the music that you choose allows the information to sink in either easily or harder. For example, listen to classical music, binary beats, natural sounds of nature as these genres of music may allow information to sink into your unconscious mind much easier.

Whenever I revise I usually listen to Mozart, as I find it easier to study with it playing in the background.

TRANSFORM

Did you know that your values and beliefs are learned from a combination of your family, friends, religion, school, geography, demographics, economics, media and significant emotional events?

- Do you know how powerful your brain is?

- Do you know how your unconscious minds absorb information from birth?

Between the ages of 0-7 your brain is like a sponge and you copy whatever you are exposed to. This is known as the 'Imprint Period'. Between the ages of 7-14 you begin to select role models, and this is known as the 'Modelling Period'. Between the ages of 14-21 you choose a group to belong to. It is fair to say that this group became your life advisory board in many ways. This is known as the 'Socialisation Period'. Between the ages of 21-35 is when you would have gathered a collection of experiences as a result of all of the above. It is believed that to some extent your 'Business Persona' resembles your first boss of authority figure. This is known as the 'Business Persona Period'.

- Have you ever heard about the 'Hierarchy of Change'? If your answer is no, then I encourage you to do some research.

Just sit back and relax and allow me to explain... well at least what I've been taught! It is human nature to protect your identity; an identity shift will change your life, because your identity forms who you are. Values determine where you spend your time. Whatever you value most you will take pride in, because it is important to you.

- Did you know that where your attention goes, energy flows and results will show?

Your beliefs are possibilities which you believe to be true. Your beliefs are 'Your Boundary Conditions'. Your beliefs change your capabilities, behaviour and the environment you choose to be in. Change your beliefs and you can change your life. Your capabilities are influenced by your behaviour. When you enhance your capabilities, your behaviour will improve, and change will happen. Your behaviour is influenced by your capabilities, and will affect the environment which you choose to be in. I hope you took that all in, take a deep breath and a moment to think about these questions:

- Where do you believe you learned some of your behaviours from?

- Who do you feel you behave like in your friendship and/or family?

Did you know that you become a combination of the five people that you spend the most time with.

The good news is, if you learned unfulfilling behaviours. You can choose to learn more fulfilling ones, right? Yes or No?

- Did you **know** that your brain processes information using representational systems after receiving data from your five senses?

These are sight, sound, touch, taste, smell and self-talk. Your brain translates the external world into an internal representation of your own unique reality, as we all have our own interpretation of reality.

- Can you remember the last movie you watched with friends or family?

ASK – BELIEVE – RECEIVE

O-kay! The kind of life you **dream** of may appear out of reach for the moment; however, I want you to take a moment to remember a week ago:

- What was you doing a week ago and how does your mind set differ **now**?

- What thoughts were you having?

- A week ago, did you k**now** that you would be reading this book and making significant changes in your life?

- How many ways will your life improve with your new expanded awareness?

Yippee! Yippee! Yippee! You are making new connections and your reality is expanding because of the k**now**ledge you have gained. You are seeing things differently **now** and its feel great... right? Yes or No? All of your memories and associated incidents are stored in your unconscious mind. Your unconscious mind is the home of your empowerment, creativity, emotions, imagination and energy. I am so excited about sharing this powerful process with you. **A**sk - **B**elieve - **R**eceive is the key to manifesting your **dream**s:

1. **Ask** - you must k**now** what you desire... You must really k**now** what you desire. The universe can't deliver your desire without

first k**now**ing what you desire to manifest into your life. Be sure to make your goal specific and clear, then it'll become attainable and near.

2. **Believe** - you have to maintain unwavering faith that what you desire will become **'yours'**. All uncertainties must be eliminated.

3. **Receive** - it is very important that you become an **action taker**. When opportunities present, you must **TAKE ALIGNED ACTION**! And do what presents; learn to be a generous giver and an excellent receiver **now**. Most importantly, you must become a vibrational match to that which you desire to welcome into your life.

This is a proven technique; sports psychologists use this technique when training athletes. It is the difference that makes the difference, it determines if **YOU WIN** or lose. Remember our minds can't determine the difference between fantasy and reality.

- Are you ready to play? Yes or No?

BY THE WAY! Repetition is crucial, like anything in life practise makes improvement and relaxation is also a key factor. Next is a brief to prepare you for the visualisation process. It'll be a good idea to either record the visualisation process on to a mobile or ask someone to read it to you.

BRIEF

Relaxation: Calm your mind and Relax your body so much so that you become at one with the visualisation process. Fully allow yourself to experience your emotions and become fully aware of your movements, breath etc taking place within your body.

Visualisation

Make the mental image so realistic that you believe that you are actually doing it **now**. You must call upon all your senses and imagine what you can see, hear, feel, smell, taste and self-talk where applicable. You must believe as you will only achieve what you deem to be true.

Language

Become totally mindful of external and internal words that are associated to your desired outcome. **Now** read each storyline until your become familiar with it, then go ahead and close your eyes to experience the visualisation process. **It's time to play...**

Visualisation Process Storyline 1

You are about to perform a serve in a badminton contest, the following components are associated: preparation, shuttlecock toss, impact of racket, ready position and shuttlecock flight then await your opponent's reaction. **Now** - take a deep breath and enhance this experience, by making the colours more vibrant, turn up the sounds which surround you (self-talk and what others are saying), notice the temperature.

See the shuttlecock in one hand and your racket in the other. Relax your shoulders and hands - feel the texture of the racket grip within the palm of your hands. Feel the texture of the shuttlecock on your fingertips, as you prepare to launch the shuttlecock into the calm air. Notice how your arms do this effortlessly, you will become aware of the shift in your body weight as your elbows and knees bend to get into position. Notice how your body is rising as you prepare to strike the shuttlecock, notice the energy in your body as you connect with the shuttlecock. See the shuttlecock as it races through the air upon impact, **now** feel the relaxation within your body as you witness your opponent attempting to return the shuttlecock and realise that they don't and your body is filled with an overwhelming desire to express your excitement and pleasure. Feels great right! - take a deep breath and just relax and preserve the positive learnings.

Visualisation Process Storyline 2
CONSCIOUSLY DESIGN YOUR DAY

Take a moment to relax...
That's right go ahead and just relax...
Deepen your relationship with yourself and the world around you right **now**. You will need to trust your unconscious and fully experience the process:

Take a deep breath and visualise tomorrow, simply say the following:

"I am taking this time to create my day, and I saturating the universe with my desires"

Now go ahead and simply visualise each phase of the day from the time you awake until the point when you go to bed. Become fully aware of what you see, hear, feel, taste, smell and self-talk. **NOW** go ahead relax, close your eyes and begin to visualise tomorrow...

Focus on all the details - in a moment not **NOW.** I want you to close your eyes and fully experience what you have created as your desired vision for tomorrow. Be sure to end your visualisation process by taking a deep breath and then blow three huge breaths onto your desired outcome for tomorrow.

Tomorrow evening simply write down your entire day and then compare the two.

- What seeds will you plant today for a brighter

OPPORTUNITIES

Throughout my life I have had many amazing unique opportunities.

I am grateful for Philip Chan telling me to apply for Channel 4 Child Genius in 2017, because not only did I have a good experience. It also contributed towards me being invited to sit test papers at the prestigious Eton College.

I am lucky to have such an amazing and inspirational family to help me create an abundance of unique opportunities. Philip Chan connected us with Mayooran Senthilmani, his wife Labosshy Mayooran and Prasanthika Mihirani to publish our first 5 books. In addition to this, if it wasn't for the support of my family and friends I wouldn't be able to achieve so much.

Throughout my life I have been fortunate enough to be nominated for a host of awards and unique opportunities.

Pastor Nicholas Nunayon and Jojus mother, encouraged my mum to contact the prestigious Eton College to pursue both my academic and sporting aspirations. Eton College was impressed by my achievements to date hat included participating in Child Genius, being an author, public speaker etc. I am so proud of myself for being chosen to be 1 of 34 boys in the entire UK (that's a ratio of 1:10000) I use this ratio as a reminder that I am able to aim high in life. When an opportunity presents, I often ask myself, how will this opportunity impact my life and the lives that I come into contact with?

I felt really honoured to have be granted this opportunity, it gave me a chance to see what life has to offer.

Philip and I delivered a 10 Seconds to Child Genius workshops to the kids at Pastor Nicholas church.

Your energy will create unique opportunities a time or set of circumstances that makes it possible to do something. So always be sure to trust yourself, by listening to your intuition and then take aligned action.

CONSEQUENCES OF DELAYED CREATIVITY

I believe that there are significant consequences of delayed creativity. As it'll have a huge negative impact on your financial education, personal development, network, experiential learning, unique opportunities, pocket money and it'll deprive you of priceless experiences that you could have had.

Creativity promotes personal expression, neuroscience and cognitive science research are constantly providing information that links creativity with intelligence; academic, social, emotional intelligence; and the development of skill sets and the highest information processing that will become increasingly valuable for children and youth.

In my view allowing children to discover their unique essence of creativity as young as possible is the difference that makes the difference.

Did you know that creative expression impacts our conceptual thinking and transfer of knowledge, judgment, recognition of relationships for symbolic conceptualisation, self-evaluation of emotions, including recognising and analysing response choices and the ability to recognize and activate memories.

Take a moment to reflect on the impact that the lack of or the abundance of creativity has had on your life?

ACTION STEPS

Take a moment to reflect on all of the content that you have been exposed to in this book and then choose three *Action Steps*

1.
2.
3.

Give each *Action Steps* a deadline date

1.
2.
3.

How will you know when you have successfully achieved each of your *Action Steps*?

1.
2.
3.

ETON COLLEGE

ETON COLLEGE

Wow! Wow! Wow!

Where on Earth do I begin?.... one thing that I must say is that it does not matter where you start, what matters most is where you choose to go.

The application process was extraordinarily phenomenal. Eton College surpassed all my expectations of a full boarding school and then some. Firstly, I was rendered speechless when I saw that the school spans over such a large area not like anything I've seen before. I can only describe it as a village as opposed to a boarding school. Different academic subject. departments are located on various roads.

How cool is that?....

Visiting the site has totally blown my mind. The staff support from start to finish were so warm, professional and supportive upon so many levels.

Somebody pinch me because I have been invited to sit papers at the prestigious Eton College, Windsor and I'm 1 of 34 boys in the entire UK (that's roughly a ratio of 1 in 100,000 boys)

Obviously just like any relationship, there are elements of privacy and confidentiality that cannot be violated, in order to uphold respect for Eton college. So, unfortunately I'll not be able to tell you about the testing procedure. All I can say is that I highly recommend that you and everyone you know do some research on what it takes for a boy to apply to Eton College, because it'll totally transform your perspective on education as it did for me, my mum, and brothers when we attended.

I can still remember our journey to Eton College, the way they greeted us, the manner in which the boys conducted themselves, the positive thoughts that were running through my mind and that warm fuzzy feeling in my tommy really made me feel like it was my home. I can remember what we had for lunch, the Q & A session, the way they treated us, the way in which they poetically spoke about education and the life time impact that they have on all the boys who become Etonians. I convey absolute gratitude for such a truly remarkable experience and memories, that I will cherish for a lifetime.

Eton College staff were amazing really kind towards all of us. They were open to answering all of our questions.

When I found out that I hadn't passed the test, I immediately realised where I had messed up. I was nervous and rushing and did not read all of the information properly. This was my first experience with academics at such a prestigious level. I didn't know how to prepare myself. As the saying goes, if you fail to plan - you're planning to fail. I felt really down because I had seen how serious they were about sports and academic studies, I really wanted to get in. what was interesting is the fact that I was 13 and only recently heard about the school, however some boys are put onto the waiting list at birth and their cut off point for applications is year 5 – so how lucky was I to have had the opportunity… finding this information out made me feel extremely proud of myself for being invited and giving it my best shot, especially due to the fact that I was raised in Eton Road and made it all the way to Eton College despite my challenges, background, class and financial background etc. so you see, never give up, because you have no idea where your efforts can lead to in the end.

My mum explained to me that everything happens for a reason, maybe it wasn't the right time for me (as I was facing serious health challenges that could have cost my life as it's a boarding school) and that I should always remember that as one door closes another shall open. It truly was a life changing experience for me and my two younger brothers. I really appreciate the fact that I was able to have such an experience with my brothers and mother right by my side to support me through the process. My little brothers were able to see what is possible for them when they choose to work smart. I see all of my opportunities as blessings and a glimpse of what is available when we choose to expand.

Did you know that Eton was founded in 1440 by King Henry VI as "Kynge's College of Our Ladye of Eton besyde Windesore” to provide free education to 70 poor boys who would then go on to King's College, Cambridge, which he founded 1441.

Eton promotes the best habits of independent thought and learning in the pursuit of excellence; provides a broadly-based education designed to enable all boys to discover their strengths, and to make the most of their talents within Eton and beyond. Eton engenders respect for individuality, difference, the importance of teamwork and the contribution that each boy makes to the life of the school and the community; supporting pastoral care that nurtures physical health, emotional maturity and spiritual richness and fostering self-confidence, enthusiasm, perseverance, tolerance, integrity and so much more.

Did you know that Prince Harry, Prince William aand Sir John Edward Nourse Heygate, 4th Baronet Heygate (19 April 1903 – 18 March 1976) was a Northern Irish journalist and novelist attended Eton College?

Engineering

EXCLUSIVE BONUS CHAPTER

There is a saying that your network equals, YOUR NET WORTH. I am certainly fortunate to be surrounded by a host of extraordinarily, phenomenal and incredible people. Therefore, I asked three of our family friends these four questions:

1. If you were to give advice to your younger self what would it be?

2. What are you really proud of?

3. What would you want to change?

4. What challenges occurred at my age (14) and what did you do to overcome them?

DR. BREMLEY W.B. LYNGDOH
(family friend)

Founder and Chief Executive of Worldview Impact Foundation

Dr. Bremley W.B. Lyngdoh is the Founder and CEO of Worldview Impact Foundation. He is a Climate Change and Sustainable Development professional with over 20 years' experience working with Governments, IGOs, NGOs and the Private Sector developing a range of innovative projects in Asia, Africa and South America aimed at producing ecologically sound and economically viable activities that contribute directly to reducing rural poverty, and generating productive sustainable livelihoods for vulnerable local communities. Bremley has all the relevant experience in project sourcing and development. His strength lies in building strategic partnerships with various governmental agencies, NGOs and multilateral development agencies. Through his previous assignments working with the United Nations and the World Bank in Asia, Africa and Latin America, he has gained expertise in the effective monitoring and evaluation of field-based programmes. Bremley developed projects on climate change adaptation, integrated agroforestry, sustainable tourism and renewable energy.

- Charity fund raising event running the Paris marathon for the Children's Trust https://www.thechildrenstrust.org.uk/

- You can support our projects by donating on http://worldviewimpact.org/causes/
- You can offset your carbon footprint by buying a rubber trees on http://worldviewimpact.com/login.php

1) If you were to give advice to your younger self what would it be?

Keep your dreams alive no matter what happens along the way and give your best each day. We need to have a powerful revolution in our hearts to create a beautiful evolution in our minds.

2) What are you really proud of?

I am very proud that Indian Prime Minister Atal Bihari Vajpayee appointed me as the Youth Representative for India during the historic United Nations Millennium Summit where I addressed world leaders during Millennium Assembly on 28th September 2000 at the UN HQ in New York.

3) What would you want to change?

If I could go back in time I would spend more time with my father in India who died from a heart attack in 2012 during his early morning run in the forest. If I had been running with him that morning I would have been able to save him and he would be still alive today.

4) What challenges occurred at my age (14) and what did you do to overcome them?

I was having challenges with balancing high school work while I was studying at St. Edmund's School in Shillong and working in my family farm to earn some pocket money. I lost my focus in time management so I joined the school musical and rebuilt my confidence by learning to perform in front of a big public audience.

JUERGEN PALLIEN
(my stock & trading mentor)

Juergen Pallien is a Master of Automation! His ability to automate success in Business and Investing creates financial liberation to pursue the most exciting and important things in life!

Juergen has used automation as the key strategy for success in his career as a serial entrepreneur, international speaker, property mogul, stock market investor and now, as a philanthropist.

Described as a Financial Liberation Genius with a Comedian's sense of humour, Juergen is on a mission to help 1 Million people achieve their own financial liberation! Educated by some of Europe's most renowned Universities in Management, Information Technology, Sales, Influence and Leadership, Juergen is fanatical about personal and professional growth and coaching. He believes that which is not growing, is dying!

Juergen is BEST known for using automation to create efficiencies in Sales, Business Growth and customer focused problem solving that led him to build and sell several companies and retire by the age of 35.

Fascinated by the stock market since he first traded in 1998, Juergen studied the investment strategies of Warren Buffett, Robert Kiyosaki and George Soros. Building on their knowledge, he has now invented several proven strategies to easily and predictably earn money without the need to sit in front of a computer all day. Juergen is committed to

sharing his stock market success formulas to help One Million people put their financial success on auto-pilot and live their passion!
When he is not helping others, Juergen likes to do sports, play chess, listen to music and read lots of books.
For your questions or to learn how you can become 1 of Juergen's million beneficiaries, contact him directly at:

- YouTube: www.youtube.com/channel/UCiFeVHbloSPO0wqWZTx8gIg/videos
- Website: www.24hprofits.com
- www.facebook.com/juergen.pallien.1
- www.twitter.com/j_pallien
- www.linkedin.com/in/jpallien

1) If you were to give advice to your younger self what would it be?

Learn to invest properly and the power of compounding. Also, to only buy luxury goods and fun stuff from the earnings of your investments rather than your net worth.

2) What are you really proud of?

The results of my students who never had any experience with the stock market but tripled their investment within in the first 10 months or were able to quit their job after just 8 months.

3) What would you want to change?

Financially liberate 1 million people in the world to enable them to live their passion and make it a better place for everyone.

4) What challenges occurred at my age (14) and what did you do to overcome them?

I was living in a very small village that limited my possibilities. I'd love to help young people on their entrepreneurial journey and provide them with the knowledge and infrastructure to make their bright ideas become a reality no matter what background they have. Everyone deserves an equal chance in life.

REGAN HILLYER
(my personal development & business mentor)

Regan Hillyer is a Serial Entrepreneur, Philanthropist, Mindset Coach and Global Speaker. She is the founder of Regan Hillyer International, a company dedicated to providing personal development and business training to men and women who have a big message they want to share with the world.

Regan specialises in helping experts uncover their true message and launch powerful personal brands, helping them make a big impact and build a legacy. Regan has trained thousands of people, helping them build multiple six and seven figure businesses location free, using powerful mindset changing tools and cutting edge business development strategies.

Regan is a certified Master of NLP, Master of Hypnosis, Time Dynamics Specialist and a Success Strategist, amongst completing many other certifications and trainings. Regan has invested in excess of half a million dollars on her own personal development and business journey and takes pride in continuously learning and growing from key industry leaders.

- www.reganhillyer.com/shop

1) If you were to give advice to your younger self what would it be?

Trust yourself. Really trust yourself. All of the answers are within you and you are able to access all of these answers at any age!

2) What arc you rcally proud of?

I'm really proud of the relationships I have in my life. My fiancé, my family relationships, my friends, my clients, my community. I am always surrounded by high vibrational people online and offline and it's taken work, but my relationships mean everything to me.

3) What would you want to change?

I'm always looking to change the number of people that I'm impacting every year.

There are so many people in the world who I want to help, and it fills me up when I am able to impact and serve at a greater level!

4) What challenges occurred at my age (14) and what did you do to overcome them?

At 14 I had no idea what I wanted to do with my life. I felt overwhelmed and confused when people asked me.

Bonus Question for my mentor:

5) What did you want to be when you grow up?

I would make up answers to try and please adults, when really, I had no idea. I went down a path of really figuring out what my purpose was in this world by diving deeper into books and personal development, and I chose to be okay with not knowing all of the answers even when people wanted me to have answers!

SURPRISE BONUS: WHAT THE EXPERTS SAY

Feng Shui to help support children, especially those having a hard time at school.

by Master Sarah McAllister - a Feng Shui consultant and horoscope specialist with over 14 years experience.

A bullied kid will be suffering from fear and lack of confidence, so you really need to make sure their home environment is supportive.

Bedroom Tips to Create a Supportive Space
If possible give children separate beds, not bunk beds as they tend to feel oppressed by either the ceiling or the top bunk. Wooden bed frames are preferable to metal ones and make sure the beds have good solid headboards. Place beds so there is a nice solid wall behind the headboard - this helps the child feel more secure. Their back is literally covered.

Study Tips
Place desks so that the chair has its back to a solid wall and child is overlooking the room or preferably looking out a window to the side too. Don't place desk against the bedroom wall to save on space, otherwise the kid is facing a brick wall, both actually and metaphorically speaking! Allow kids to study at kitchen table, as sometimes they just need to have people nearby in order to concentrate.

Respect their gut instincts
Little people have very good instincts and will know what colour they want in their room or where they want to sit to study, respect this always.

Put their artwork and pictures of friends on the bedroom wall
It sounds obvious, but some parents overlook this and leave a child's room bare and austere. Kids (and adults) **LOVE** to see images of their friends and also symbols of their success around them. Ideally place these pictures above eye level, as looking upwards activates vision and aspiration, whereas looking downwards concentrates us in the past.

Keep Electronics at a minimum in their bedroom
Electrical alarm clocks, cordless phones, mobile phones, computers - all must be either switched off or kept out of the bedroom so that the energy can settle during the night time and not interfere with the delicate bioelectrical field of your child which is still developing.

Use Natural Paints & Organic Materials if possible
Indoor air can be more polluted than outdoor air, so a good air filter is recommended, and use of natural paints. Chemical paints and artificial fabrics can create allergies in your child. The last thing they need is a skin condition or unsightly reactions due to allergies when they are already feeling vulnerable at school.

Use Chinese Horoscope Wisdom to help your child

Ask us to 'open the horoscope' of your child - we can decipher whether there is the presence of unhelpful authority (bullying) in the chart and advise colours to wear (as underwear if they have to wear a uniform), little pictures or totem animals to carry with them to support their chi, what foods are helpful to them and which directions within the house are best to occupy. *We had one mother from Croydon concerned with the amount of tickings off her son was getting at school, and my students and I went around to do a case study, moved his bed to a more empowering position and lo and almost overnight there was no more trouble. I also helped a kiddie in Holland Park to sleep through the night (previously waking up at 4am and bouncing into the parents room!) just by moving her bed to a more supportive area of the room and aligning it to 'quiet' energy as opposed to 'active' energy.*

Another young boy with Attention Deficit Disorder was immediately helped by moving the position of his bed and performing a space clearing - the father was initially sceptical but developed a respect for what I had done based on the results he had witnessed.

This type of refined, intricate and powerful classical Feng Shui is not a self-help subject - the above pointers are useful basics, but don't substitute a professional Feng Shui consultation and horoscope analysis. For further information please visit www.myfengshuifriend.com

email info@fengshuiagency.com

I URGE YOU TO DREAM BIG

Having **dreams** can make you feel happy. The more important that your **dreams** are to you, the more you will want to hold on to it!!

I want you to take a moment to think about what you would do if you weren't afraid to have **FUN**, **DREAM BIG** and make **money** doing what you **LOVE**? That's right... Just go ahead and allow yourself to feel your deepest and wildest **dreams** deep within your heart. After all we all have **dreams**... right? Yes or No? Fear simply holds you back. Remember you are a ***creator***, when you choose to move beyond your fears you can begin to feel free.

If you don't change you can become extinct, let's face it... change is always happening. Each and every day visualise your **dreams** as often as possible... imagine yourself enjoying your **dreams** because it can lead you to it. Movement in a new direction can help you achieve your **dreams**... so don't worry too much about being different to your peers. Difference is also good, so follow your own **dreams** and remain true to yourself. The quicker you let go of peer pressure and your old negative values and beliefs, the sooner you can find new positive values and beliefs that will serve you well. When your values and beliefs change so will your life. It will serve you best to pursue your **dream**s instead of simply living your life in default mode following your peers etc. We both know that holding onto old negative values and beliefs will not lead you to new positive results. When you realise that you can achieve your **dream**s, your life will begin to transform if you choose to. Acknowledging small change early can help you to adept to **big**ger changes that are to come, simply allow your **dream**s to flow and then you can begin your dance with the universe.

Sabrina Ben Salmi BSc

WORDS FROM THE HEART OF A GRANDMOTHER

Parents: You have to help our children and youth, because they are tomorrow's leaders. If your child has a **dream,** you must help them to nurture that seed for it to grow. Without TLC that seed cannot grow and your Childs' **dream** will surely die. It is not about pushing or pressuring your child. Ask your child how you can be of assistance and best support them to grow their **dream**? If you notice that your child is reaching out for support with a desire to be creative. Encourage them in each and every way, but always remember that it is your child's **dream** not yours. Also, too much pressure could disrupt their creative process. Take one step at a time and enjoy the experience with them. Simply embrace this priceless experience together and give them a little independence as this will help them to learn valuable life lessons along the way.

Children and Youth: If you have a **dream** don't let anyone take your **dream** away from you because that **dream** is yours and it was created by you and no one but you has the ability to breathe life into your **dream**. Always believe in yourself and your **dreams** 100%. There is no such thing as too young; even if you are 5 years old give it your all. In the end at least you will know that you have at least tried your best.
Mary Paul

(((CONGRATULATIONS)))

Trust yourself, as you k**now so** more than you think you do. Simply inhale inspiration and exhale action. **G**o ahead and do what you k**now** you ought to do. The answer you have always been waiting for is yes you can, do, be and have whatever your heart desires.

It is an absolute honour to welcome you to our family. What are you going to do to celebrate your achievements?

ABOUT THE AUTHOR
AS SEEN OF TV, RADIO & NEWSPAPERS

Tray-Sean Ben Salmi aka I'm That KID is not your average 14yr old. Tray-Sean Ben Salmi is a 14yr old Amazon No.1 Award Winning Author, Public Speaker Award Winning, Presented award for TruLittle Heros Award 2018, Guest Speaker at The Beat You Expo, Multi-award winning child advocate, Made For Mums Judge 2018, former member of Team Trouble (participated in campaigns for Sainsburys, Legoland, Warner Bros, Sony and Made For Mums to name a few) founded by Shadia Daho, Amazing Arabella & JD The Super Car Kid, Child Genius 2017 1 of 20 smartest children in the UK, 1 of 34 boys invited to sit at the prestigious Eton College for Boys. Official Judge for Made For Mums Toy Awards 2018 via Team Trouble, An award winning author of Kidz That Dream Big, Former Radio Show host, Regan Hillyer International Be Your Brand Fellow, Author of 10 Seconds To Child Genius, Winner of TruLittle Heros Award - Academic 2017, Public speaker, a business/personal developments mentor & coach and founder of I'm That KID Blueprint covers:

- I'm That KID - Bridging the Gap Between Fathers & Sons
- I'm That KID – Creating A Vision Board for My Future
- I'm That KID – Taking To The Stage
- I'm That KID - Inspiring My Community To Pay It Forward
- I'm That KID - There's A Book Inside ME
- I'm That KID - Families That Play Together, Stay Together

- I'm That KID - Empowering You To Step Into Your POWER
- I'm That KID - BEING The Change That I Desires To See In The World

And co-founder of 10 Seconds To Child Genius who is here to help child to plant the seed to create a brighter Future. Tray-Sean's signature program: I'M That KID Blueprint™

BEN SALMI FAMILY MANTRA

"BEN SALMI TEAMWORK, MAKES THE DREAMWORK

We believe that there is no such thing as failure only feedback.

We also believe that the journey of one thousand miles begins with a single step in the right direction

FAMILY ANTHEM

If you want to be somebody,
If you want to go somewhere,
You better wake up and PAY ATTENTION
I'm ready to be somebody,
I'm ready to go somewhere,
I'm ready to wake up and PAY ATTENTION!

The question is *ARE* ***YOU?***

ABOUT TRAY-SEAN'S SIBLINGS

AS SEEN OF TV, RADIO & NEWSPAPERS

Lashai Ben Salmi aka DREAMPRENEUR is not your average 18yr old. She is a multi-award winning Youth Advocate, Presented award for TruLittle Heros Award 2018, Content Creator for The Korean Cultural Centre, Winner of TruLittle Heros Award - Entrepreneur 2017, Speaker at Virgin Money Lounge Historical Black History Month first ever event, Guest Speaker at The Beat You Expo, Guest Speaker at Mercedes Benz World 10th April 2018, High Profile Club, YouTuber with 25K plus subscribers and over 4M plus views (Korean Channel), An award winning author of Kidz That Dream Big, Andy Harrington ACE Coach, Former International Radio Show host, Winner of Regan Hillyer International Scholarship, a speaker, a business/personal developments mentor & coach, founder of Blossom Tree Photography & Videography (Produced content for Shadia Daho for Amazing Arabella, JD The Super Kid & Team Trouble in association with Legoland Resort, Harry Potter, Little Mix and Disney Pixar, Sony, Warner Brothers & Universal etc.) co-founder of A Precipice of A Dream and founder of Put The RED Card Up To bullying & My Journey - Giving Youth Several Reasons to Smile who is here to help children and youth to plant the seed for an abundance of unique opportunities via a variety of products and services to assist you to create a brighter future

Lashai has been mentored by some of the leading name within the personal development world Regan Hillyer, Andy Harrington, Cheryl Chapman, Harry Singha,

Ralph Plumb, Sammy Blindell to name a few. Lashai has shared the stage with the likes of the late Dr Miles Monrune, Dr John Demartini, Andy Harrington, Robert G Allen and Ralph Plumb to name a few.

If you are looking for an inspiring, wise, talented, refreshing and powerful speaker then 18yr old Lashai Ben Salmi is guaranteed to make a big impact at your event. Lashai has been a part of the personal development world since the age of 11yrs. Lashai has a burning desire to transform lives with her stage presence, knowledge and wisdom! Lashai's signature topics include: Congruency, Alignment, Self-Belief, YouTube, Social Media, Connection, Inspiration and Motivation.

Lashai's signature program: The Stepping Stone's Formula™

Book: Kidz That Dream Big: Dreams Do Come True https://www.amazon.co.uk/dp/1912547066/ref=cm_sw_r_cp_api_mwbUAbS8BTQHE

Facebook page: Kidz That Dream Big: https://www.facebook.com/Kidz-that-Dream-BIG-154694734627138/

AS SEEN OF TV, RADIO & NEWSPAPERS

Yasmine Ben Salmi is not your average 11yr old. Yasmine Ben Salmi aka LovePreneur is an 11yr award winning author of The Choice is Your - 10 Keys Principles To Create A Happier Lifestyle, Winner of TruLittle Heros Award - Creative 2017, Guest Speaker at The Beat You Expo, Former International Radio Show Host, Member of Team Trouble (participated in campaigns for Sainsburys, Legoland, Warner Bros,

Sony and Made For Mums to name a few) founded by Amazing Arabella & JD The Super Car Kid, Yasmine's signature program: Your Thinking C.A.P For Living & Loving Life™, Yasmine's Dog Walking Service "Woof-Woof your dog is here", Nominated for a R.E.E.B.A Award 2017, Winner of Radio Works Authors Awards 2017, Nominated for National Diversity Award 2017, founder of Mother and Daughter Connect Collection and founder of Lovepreneur.

Book: The Choice Is Yours: 10 Key Principles to Create a Happier Lifestyle https://www.amazon.co.uk/dp/1912547082/ref=cm_sw_r_cp_api_JbaUAbR7K3MNS

Facebook page: Lovepreneur: https://m.facebook.com/YasmineBenSalmiakaLovePrenur/

AS SEEN OF TV, RADIO & NEWSPAPERS

9yr old Paolo Ben Salmi aka Pint Size Adventurer is not is not your average 9yr old. Paolo is Water-to-Go's youngest ever ambassador! https://www.watertogo.eu/paolobensalmi blog about Water-to-Go and Paolo: https://www.watertogo.eu/blog/meet-paolo-water-to-gos-youngest-ever-ambassador/ Paolo Ben Salmi is an award winning author of Pint Size Adventurer - 10 Keys Principles To Get Your KIDS off their iPads & Into The Wild, Award Winning Public Speaker (who has spoken at eleventh such as Mercedes Benz World), Former member of Team Trouble (participated in campaigns for Sainsburys, Legoland, Warner Bros, Sony and Made For Mums to name a few) founded by Amazing Arabella & JD The Super Car Kid, Paolo's signature program is called The Abundant Adventure Inventor™, 2nd place in TruLittle Heros Award - U12 Entrepreneur 2017, Guest Speaker at The Beat You Expo, Mercedes Benz World 10th April, Official Judge for Made For Mums Toy Awards 2018 via Team Trouble, Former International Radio Show host, 229/17 Paolo made history by being the youngest to interview Dr John Demartini: https://www.facebook.com/350400542063654/videos/363072487463126/, personal developments coach and founder of Pint Size Adventurer who is here to help you to plant the seed toward self-discovery, exploration of the internal and external world and adventurer in abundance via a variety of products and services to assist you to create a brighter future

Book: Pint Size Adventurer: 10 Key Principles to Get Your KIDS off Their iPads & Into the Wild https://www.amazon.co.uk/dp/1912547031/ref=cm_sw_r_cp_api_iwXXAbMZRM7QA

Facebook page: Pint Size Adventurer: https://m.facebook.com/paolobensalmiakapintsizeadventurer/

AS SEEN OF TV, RADIO & NEWSPAPERS

Amire Ben Salmi is not your average 5yr old. Amire Ben Salmi aka Mr. Intelligent is a 5yr old award winning author of Because I AM Intelligent - 365 Affirmations To Brighten Up Your Day, Guest Speaker at The Beat You Expo, Member of Team Trouble (participated in campaigns for Sainsburys, Legoland, Warner Bros and Sony to name a few) founded by Amazing Arabella & JD The Super Car Kid, Amire's signature program is Easy-As-P.I.E Affirmatios™ and He's founder of

Because I AM Intelligent who is here to help you to plant the seed toward having fun learning during childhood, Positive Affirmations, Fun and Creativity in abundance via a variety of products such as a book with a matching colour car and 52 affirmation cards to assist you to create a brighter future

Book: Because I AM Intelligent 365 Affirmations to Brighten Up Your Day
https://www.amazon.co.uk/dp/1912547023/ref=cm_sw_r_cp_api_gcaUAb6A5W5SJ

Facebook page: Because I AM Intelligent:
https://m.facebook.com/BecauseIAMIntelligent/

KEEP IN TOUCH VIA SOCIAL MEDIA

FACEBOOK: I'M THAT KID
TWITTER: @traybensalmi
INSTAGRAM: authortrayseanbensalmi
LINKEDIN: Tray-Sean Ben Salmi

Please do not hesitate to contact me

YOUNG CITIZEN

Nurturing the special bond of fathers and sons

Affordable Carpet

SURPRISE BONUS

What are you good at?

What do you enjoy doing?

What are you not good at?

What do not enjoy doing?

What vision do you have for your future?

What fears to you have about your future?

What positive beliefs can you choose to overcome your fears?

List 10 affirmations that makes you feel excited about your future.

What version of you do you need to become in order to fulfil your desires?

Star of the week
Tray-Sean Ben Salmi
Help

Salford

Virgin money

Luster's PRODUCTS INC.

amazon.com BEST SELLING AUTHOR
10 SECONDS TO CHILD GENIUS
AS SEEN ON TV, RADIO, NEWSPAPERS AND MAGAZINES
TRAY-SEAN BEN SALMI
&
PHILIP CHAN
"10-Seconds Maths Expert"
Awards Winning Authors
Lone Parenthood
Sabrina Ben Salmi BSc
DREAMING
BIG TOGETHER
Mamas Secret Recipe™
Sabrina Ben Salmi BSc

PINT SIZE
ADVENTURER
10 Key Principles
To Get Your KIDS
off Their iPads
& Into The Wild
PAOLO BEN SALMI

Because
I AM Intelligent
365 Positive Affirmations
To Brighten Up Your Day
Amire Ben Salmi

amazon.com
BEST
SELLING
AUTHOR

THE GAME
CHANGER

Made in the USA
Monee, IL
04 November 2026

67380408R10059